When You See Me

The Earth's Story

By L.J. Hill

Illustrated by TullipStudio

Published by EarthVille Publishing Printed in the United States of America

First Edition

To the children of today and of the future
who will inherit this great planet.
May you explore its wonders and protect
its beauty.

L.J Hill

This Book Belongs To

- - - - - - - - - - - - -

When you see me, I'm big and round,

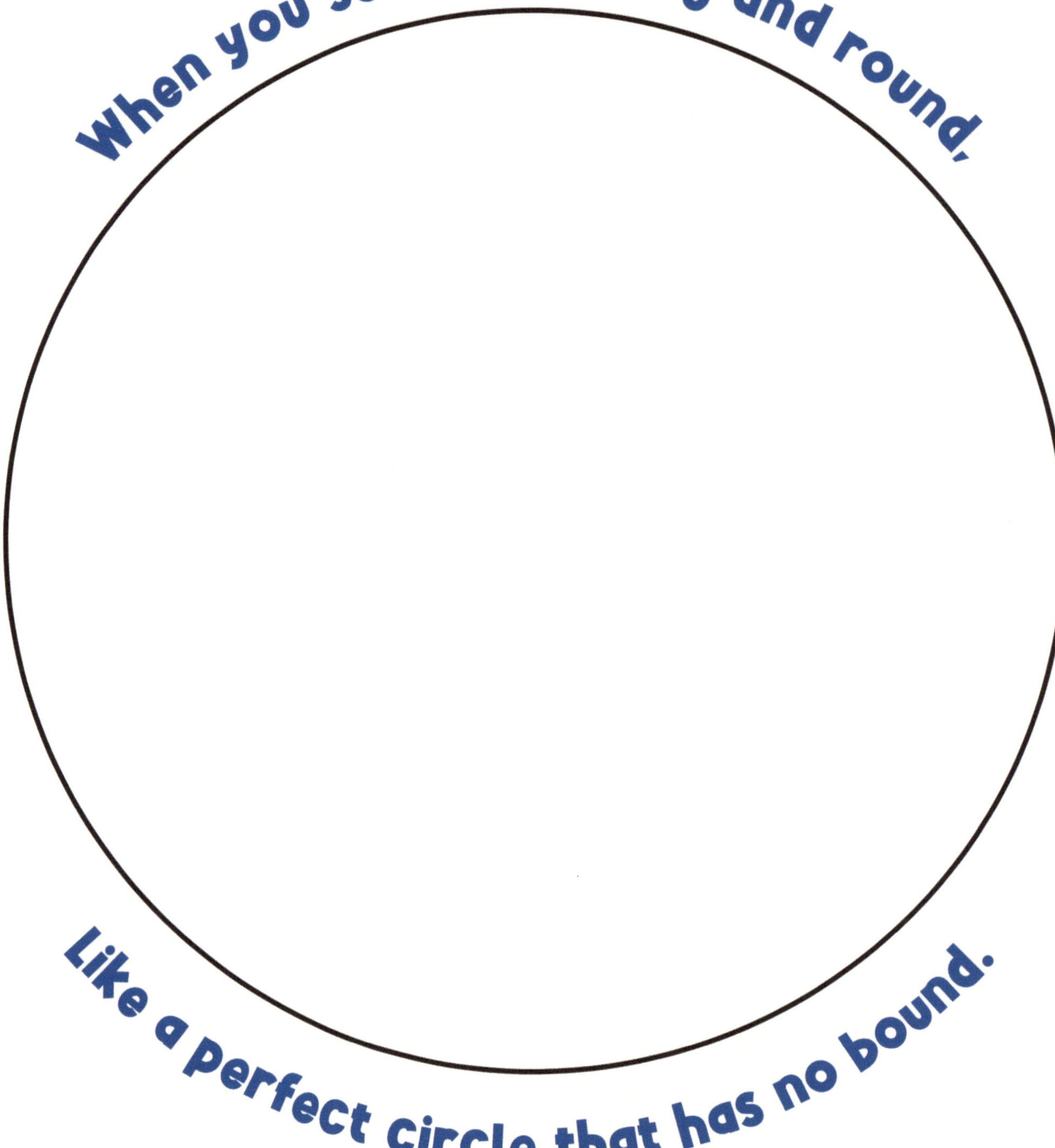

Like a perfect circle that has no bound.

Like a bouncing ball,
hopping all around, moving in every
direction it goes up and down.

Like tiny little bubbles
floating with grace, twirling and swirling,
filling empty space.

I'm big and round,
you just wait and see!

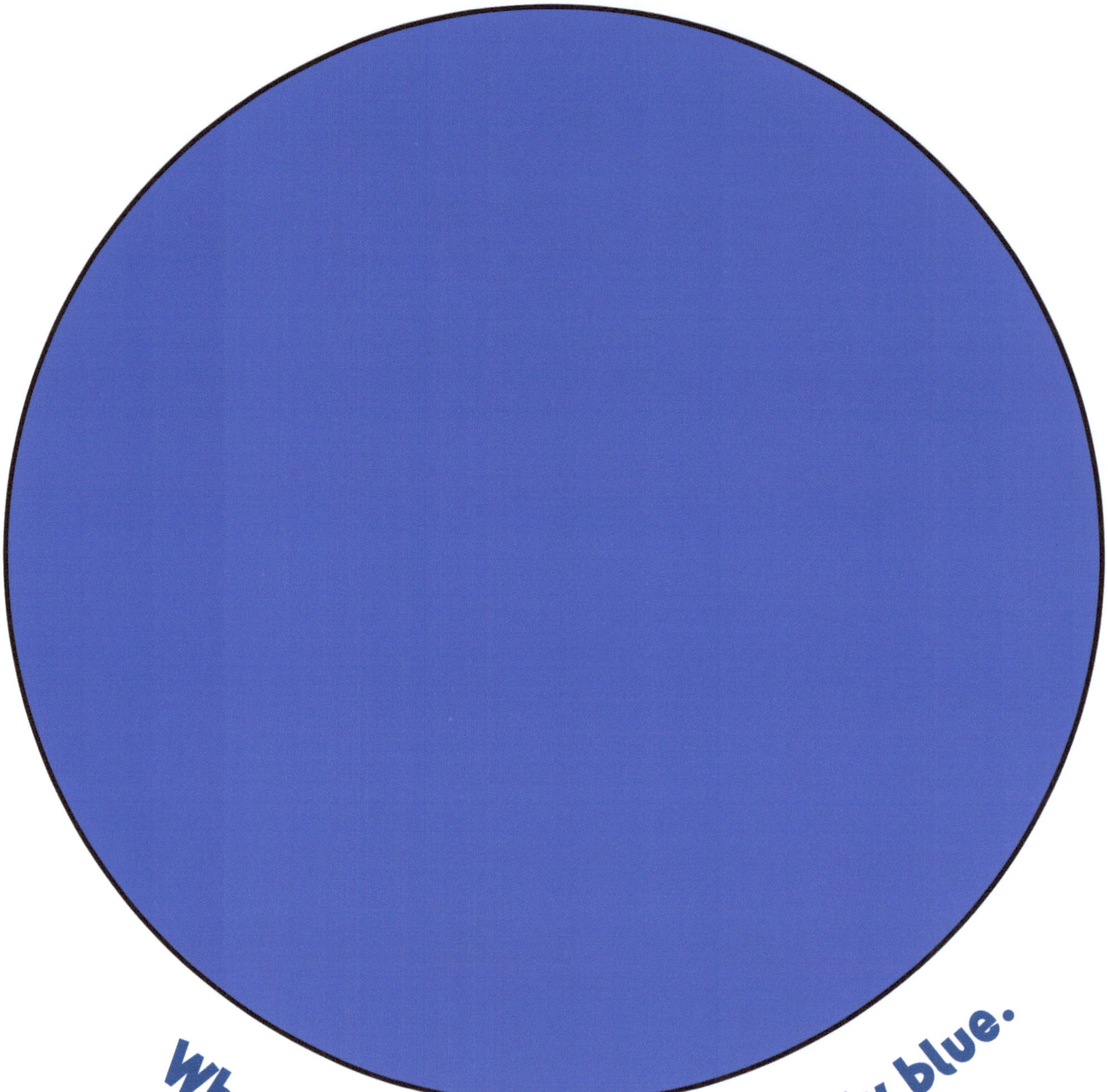

When you see me, I'm mostly blue.

Like the blue sky stretching far and wide,
touching every part of the world,
for everyone to see.

Like the blue bird gliding in the air,
living freely in a world without a care.

Or just like the never-ending blue ocean water that sparkles and glows, with a touch of sunlight as they gracefully flow.

Oh, what a beautiful color I am, I'm round and blue, you just wait and see!

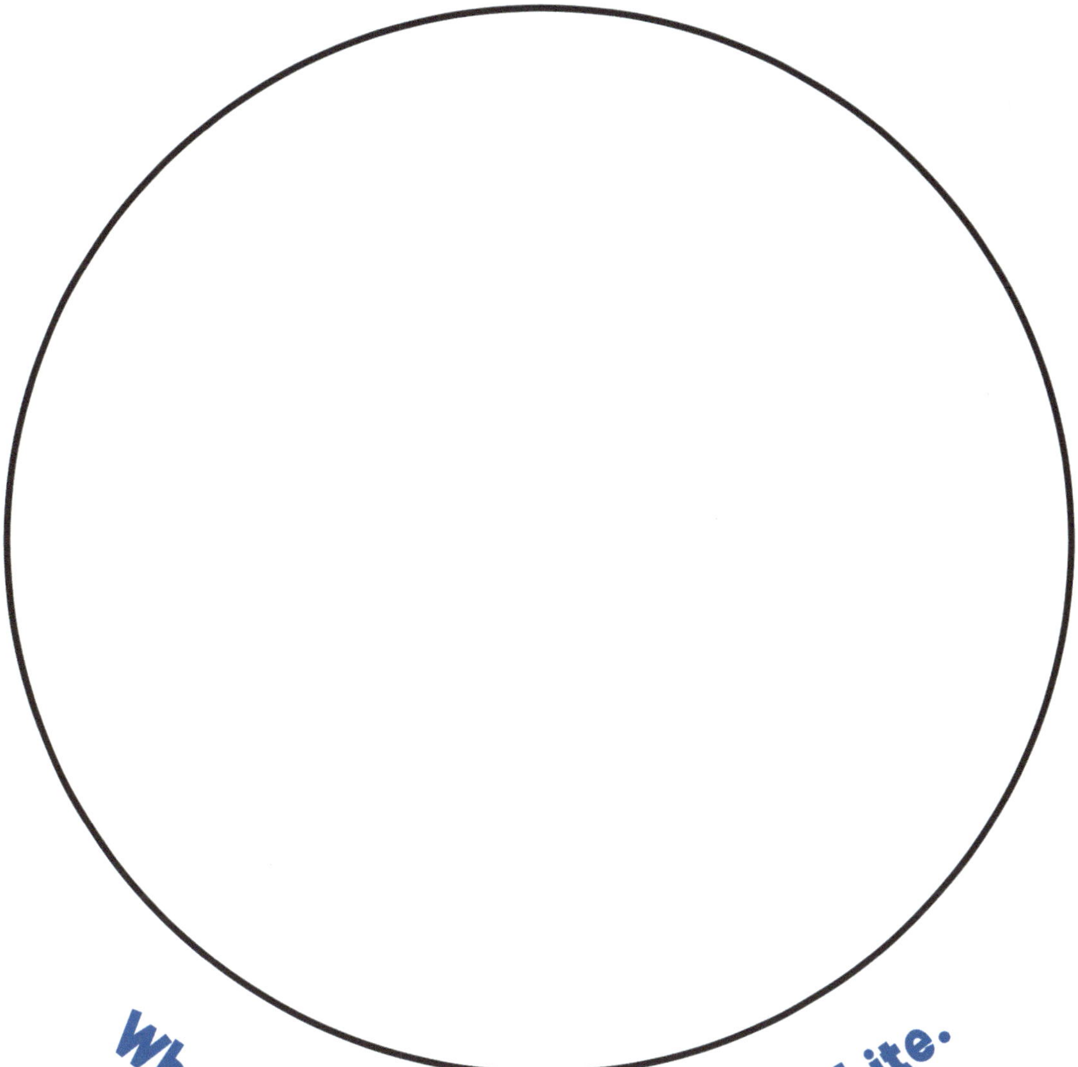

When you see me, I'm also white.

Like white stars, that sparkle in the dark, looking like little tiny diamonds, twinkling and winkling to say goodnight.

Like big fluffy white clouds, so soft and light, floating in the air making all kinds of shapes. They make us wonder as they pass us by, where are they going? Are they leaving us to hide?

Or just like the pure white snow that's gently falling, making beautiful snowflakes big and small.

Oh, what beautiful colors I am!

I'm big, round, blue and white,
you just wait and see!

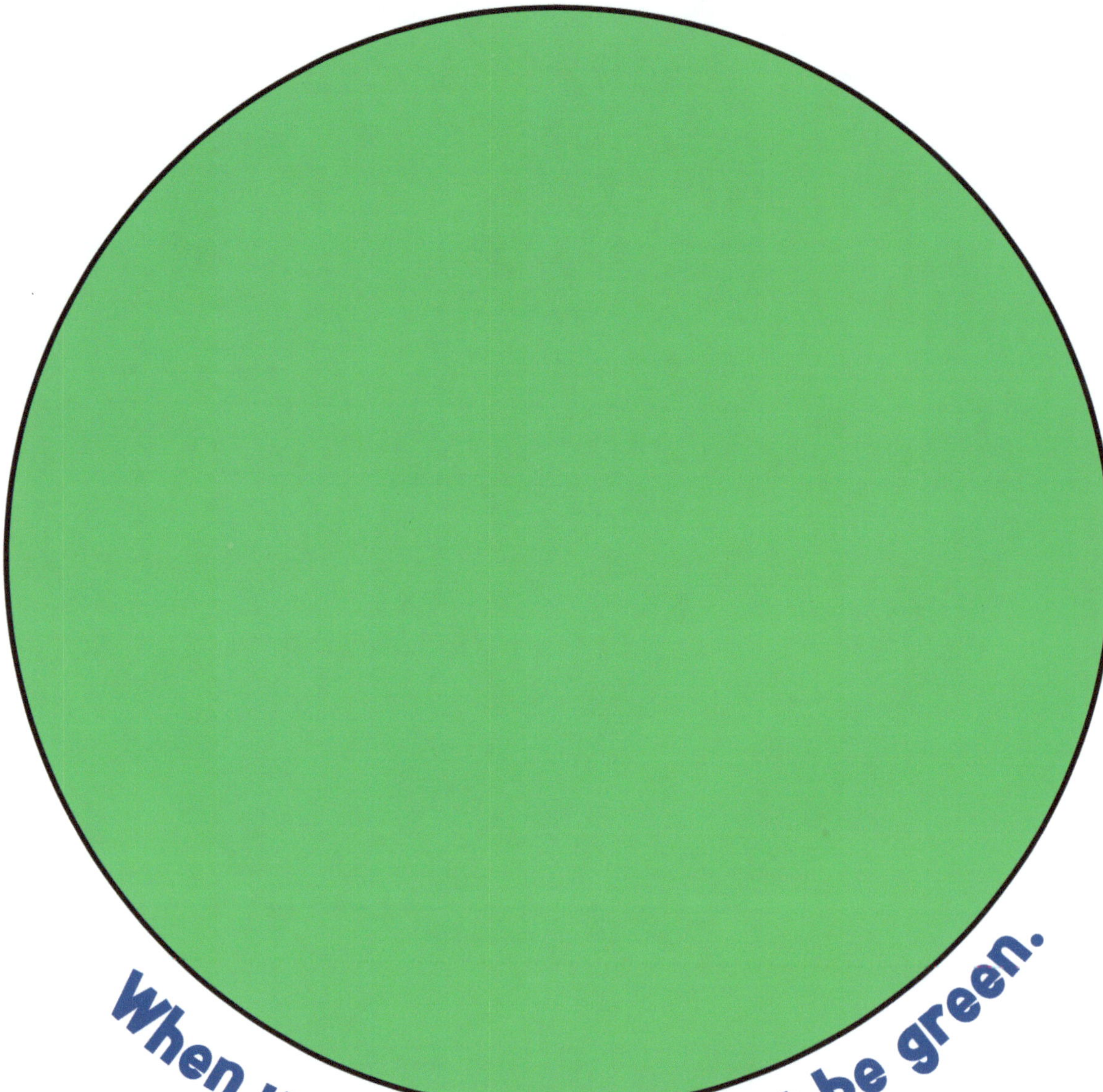

When you see me, I might be green.

Like a little green grasshopper,
hopping on its feet, leaping and bounding
with a rhythm so sweet.

Like the tall green grass, standing up high,
covering the ground as far as
the eye can see, inviting us to run and
play, it's everywhere, just look around.

Or just like the green leaves you see on plants and trees, swaying in the breeze so lovely and free.

Oh, what beautiful colors I am!
I'm big, round, blue and white and also
green, you just wait and see!

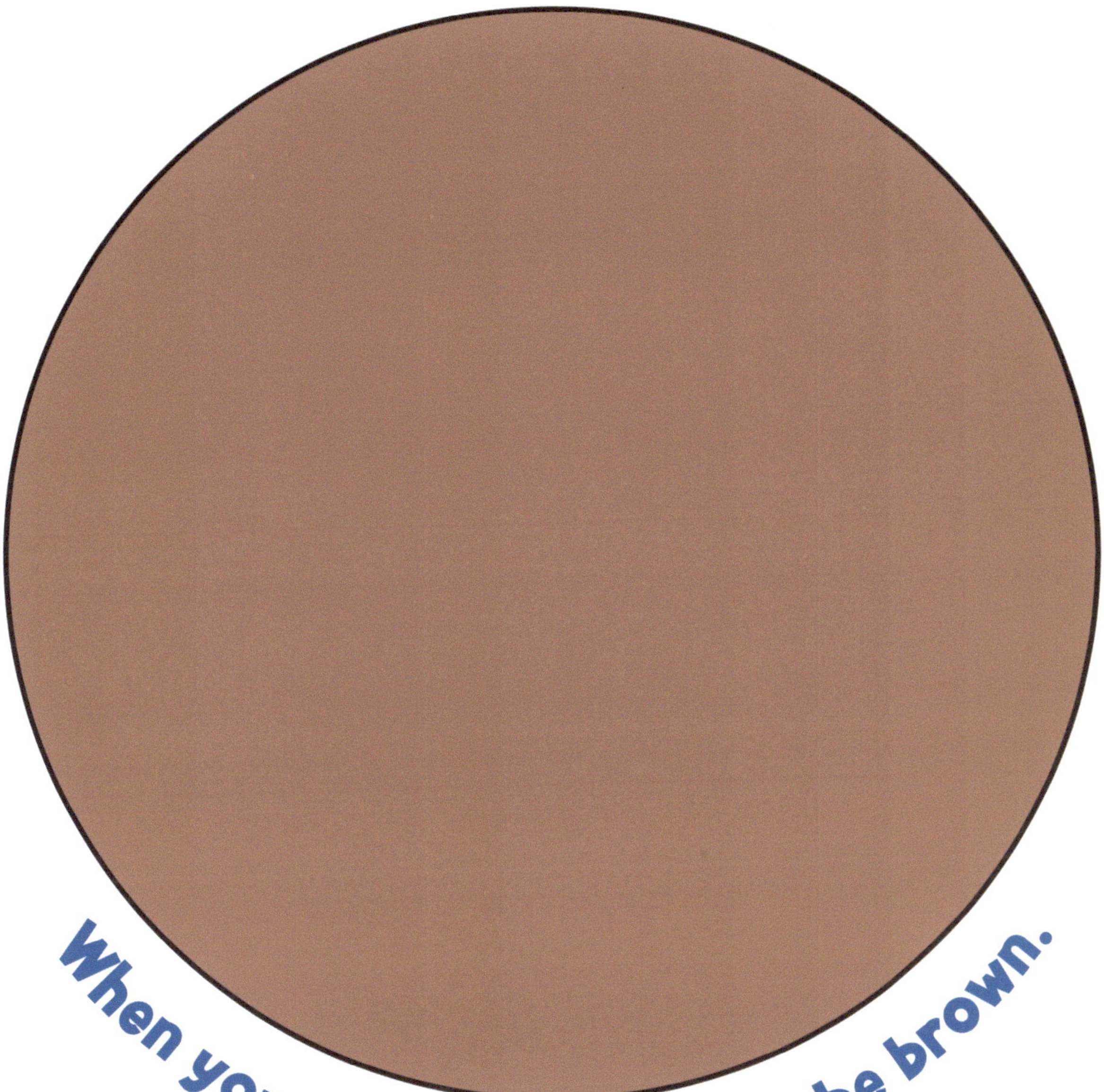

When you see me, I can also be brown.

Like a shiny brown penny,
you might find on the ground.

Like a hairy brown bear, roaming through nature, exploring the world without a single care.

Or just like the mighty brown land made of soil and rocks, together they form to make mountain tops.

I'm big, round, with colors of blue, white, green and brown, you just wait and see!

When you see me, I'm all these things,
together they make me,

Your Home,
Your Earth A place we all share!

www.ingramcontent.com/pod-product-compliance
Lightning Source LLC
LaVergne TN
LVHW071240210926
848486LV00007B/313